I0813685

SPACE SYSTEMS
STARS AND THE SOLAR SYSTEM

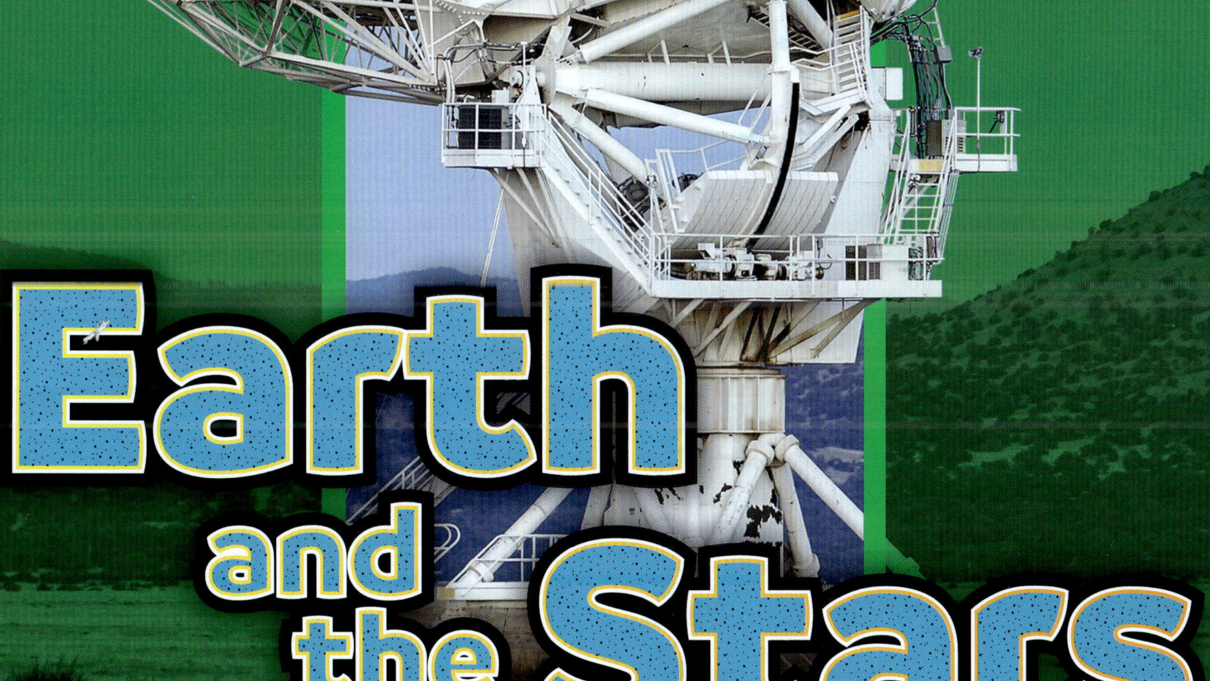

Earth and the Stars

Blaine Wiseman

LIGHTBOX
openlightbox.com

LIGHTBOX

Go to
www.openlightbox.com
and enter this book's
unique code.

ACCESS CODE

LBXH4425

Lightbox is an all-inclusive digital solution for the teaching and learning of curriculum topics in an original, groundbreaking way. Lightbox is based on National Curriculum Standards.

STANDARD FEATURES OF LIGHTBOX

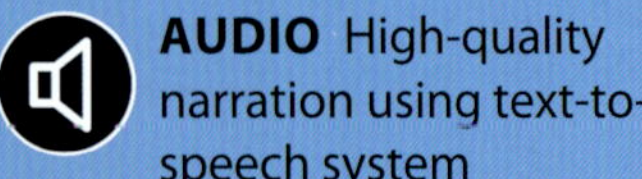

AUDIO High-quality narration using text-to-speech system

ACTIVITIES Printable PDFs that can be emailed and graded

SLIDESHOWS Pictorial overviews of key concepts

VIDEOS Embedded high-definition video clips

WEBLINKS Curated links to external, child-safe resources

TRANSPARENCIES Step-by-step layering of maps, diagrams, charts, and timelines

INTERACTIVE MAPS Interactive maps and aerial satellite imagery

QUIZZES Ten multiple choice questions that are automatically graded and emailed for teacher assessment

KEY WORDS Matching key concepts to their definitions

Earth and the Stars

SPACE SYSTEMS
STARS AND THE SOLAR SYSTEM

Contents

The Stars: Our Guiding Lights

On a clear, dark night, the sky seems full of stars. These tiny lights do more than just twinkle in the distance. For thousands of years, people have used the stars to help understand their place in the universe. Ancient travelers used the stars as a map to help them **navigate**. Modern **astronomers** study stars to map the universe.

No one knows exactly how many stars there are. The stars that can be seen with the **naked eye** are a tiny fraction of the total number. Astronomers believe there are billions of trillions of stars, maybe even more.

The closest star to Earth is the Sun. Compared to other stars, the Sun is only average-sized. However, because it is so close to Earth, its light blocks other stars from being seen during the day. Stars are much bigger than they appear. Stars are made of burning gas. They look small because they are so far away. The light people see from stars is fire. These fires are so large, and burn so bright and hot, that some stars can be seen from thousands of **light years** away.

There are **9,096 stars visible** to the **naked eye**.

The **furthest** star visible to the **naked eye** is called **V762 Cassiopeiae**.

Patterns of Movement

Every night, the stars rise in the east and set in the west. It looks like they are moving across the sky, but it is actually Earth that is moving. Earth rotates once per day, which is why people see the Sun, Moon, and stars move across the sky each day. Earth is also moving in **orbit**. This changes the way people view the night sky over a longer period of time than rotation.

As Earth moves around the Sun, people see stars from a changing angle. This changing angle is called **parallax**. It can make some stars appear to move and others seem to stay still. While Earth's rotation makes the stars rise and set each night, parallax motion takes a year to repeat.

Over a long period of time, people can even see the movement of the stars themselves. Proper motion is the permanent movement of stars from one part of the sky to another. It is difficult to see stars move without the help of telescopes and other tools. Stars are so far away that it can take people many years to notice that a star has moved even a tiny amount. It may seem like a small distance, but a star must move millions of miles through space before people notice.

Astronomers can use parallax to determine how far a star is from Earth by measuring how much it appears to move.

Parallax in Motion

Imagine your head is Earth and your thumbs are stars. Hold your thumbs in front of your face with one further away than the other. Move your head from side to side. Your closer thumb appears to move much more than your further thumb. This is an example of parallax.

Stellar Timeline

Stars have been a source of mystery and wonder since the first human beings looked up at the night sky. Curious people have found new ways of seeing the stars. As understanding has grown, new discoveries have been made.

1006 AD

A **supernova** known as SN1006 is observed by astronomers in Asia, the Middle East, and Europe. It is the brightest **stellar** event in recorded history.

About 130 BC

Ancient Greek astronomer Hipparchus creates the first known star catalog. A copy was found in an Italian museum in 2005, as part of a globe held by a 2,000-year-old statue. Every other copy was lost thousands of years ago.

1838

Friedrich Wilhelm Bessel becomes the first person to measure the distance to a star. He used parallax to measure the distance to 61 Cygni. It took him 28 years of research and work to finally conclude that 61 Cygni is about 10.3 light years from Earth.

2020

Astronomers using a telescope in Chile discover a star that seems to dance around a **black hole**. The movement of the star further confirms Albert Einstein's theory of general relativity. Einstein's theory suggests that gravity bends space, pulling lighter objects toward the center of a **mass**.

1718

British astronomer Edmond Halley discovers proper motion of stars by comparing his calculations to another astronomer's from 1,800 years earlier. He observes that some stars had moved further than the stars around them.

1901

American astronomer Annie Jump Cannon creates a new system for organizing stars. She makes a catalog of 1,122 stars, organized by temperature. By 1924, she had used the system to classify more than 225,000 stars.

Gravity and the Stars

Gravity is one of the most powerful forces in the universe. The planets in the solar system orbit the Sun because of gravity. On Earth, gravity keeps things on the ground and makes objects fall down. Understanding gravity helps astronomers learn about distant objects in space. It can even tell them how much a star weighs. The heavier something is, the stronger its gravity. Once astronomers know the strength of a star's gravity, they can estimate its mass.

A star does not have a solid surface, or ground, like Earth does. The burning gases that make up a star's surface rise and fall. Burning balls of gas that rise from the surface are called granules. Astronomers measure the time it takes for a granule to rise and then fall back to the surface. The speed of this process tells them the strength of gravity on the surface of the star.

The Life of a Star

Each star goes through different stages over time.

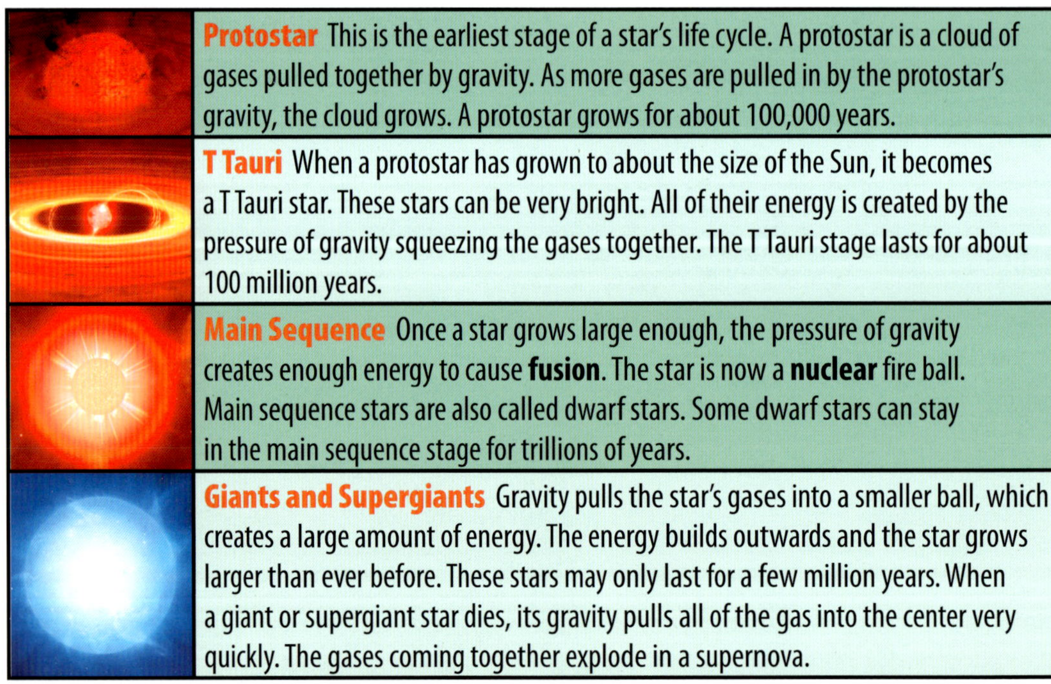

Protostar	This is the earliest stage of a star's life cycle. A protostar is a cloud of gases pulled together by gravity. As more gases are pulled in by the protostar's gravity, the cloud grows. A protostar grows for about 100,000 years.
T Tauri	When a protostar has grown to about the size of the Sun, it becomes a T Tauri star. These stars can be very bright. All of their energy is created by the pressure of gravity squeezing the gases together. The T Tauri stage lasts for about 100 million years.
Main Sequence	Once a star grows large enough, the pressure of gravity creates enough energy to cause **fusion**. The star is now a **nuclear** fire ball. Main sequence stars are also called dwarf stars. Some dwarf stars can stay in the main sequence stage for trillions of years.
Giants and Supergiants	Gravity pulls the star's gases into a smaller ball, which creates a large amount of energy. The energy builds outwards and the star grows larger than ever before. These stars may only last for a few million years. When a giant or supergiant star dies, its gravity pulls all of the gas into the center very quickly. The gases coming together explode in a supernova.

About 90 percent of stars, including the Sun, are main sequence stars.

Observing the Stars

Early astronomers gave names to the stars they could see with the naked eye. They grouped the stars into **constellations**. Centuries later, astronomers developed the telescope. This gave humans a new, closer view of the stars. People could even observe stars that had never been seen before.

Today, astronomers have developed high-powered telescopes and other stargazing technologies. The Hubble Space Telescope is a powerful telescope and observatory that was launched into space in 1990. Being in space gives it a closer, clearer view of the stars than Earth telescopes. Hubble is able to see objects ten billion times fainter than those the human eye can see. One of Hubble's most famous photos is the "Pillars of Creation." This image captures the process of stars being created in a distant **nebula**. Hubble has also taken detailed photos of a supernova.

Measure the Distance Between Stars

The distance between objects in the sky is measured in degrees. Astronomers use technical instruments to find the angle between stars. Luckily, you can use your hand to estimate stellar angles.

1. Stretch out your thumb and little finger as far as they can go. The angle between them is about 25 degrees.
2. Stretch out your index finger and little finger. This angle is about 15 degrees.
3. Close your fist. The width of your fist is about 10 degrees.
4. Hold your index, middle, and ring fingers together. Their width is about 5 degrees.
5. Hold out your little finger. The width of your fingertip is about 1 degree.

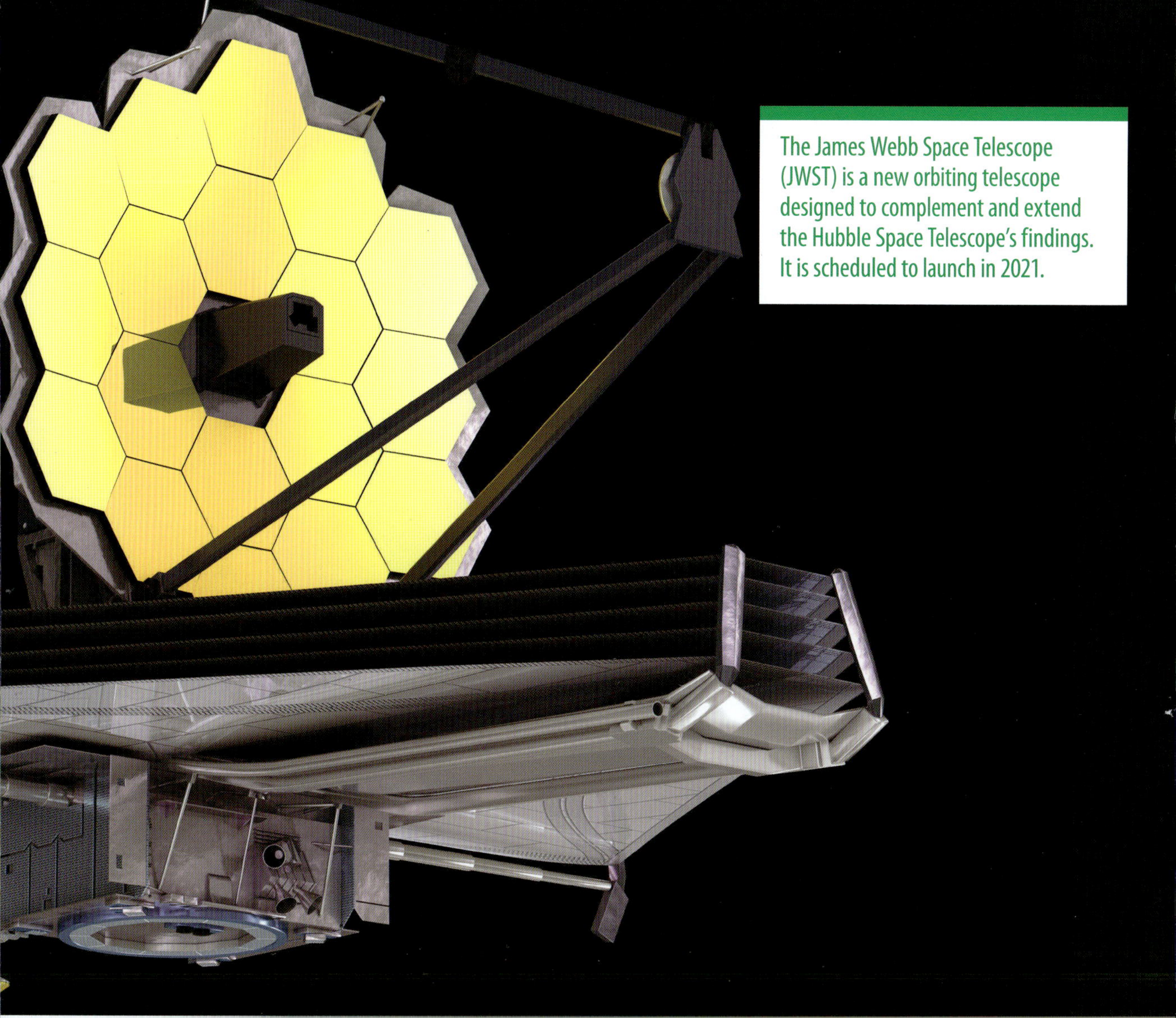

The James Webb Space Telescope (JWST) is a new orbiting telescope designed to complement and extend the Hubble Space Telescope's findings. It is scheduled to launch in 2021.

Steps:

1. Look up into the night sky and choose two stars.
2. Hold your hand at arm's length and use your fingers to estimate the angle between the two stars.
3. Write down your results.
4. Compare your own results with those of a friend, family member, or classmate measuring the same two stars.

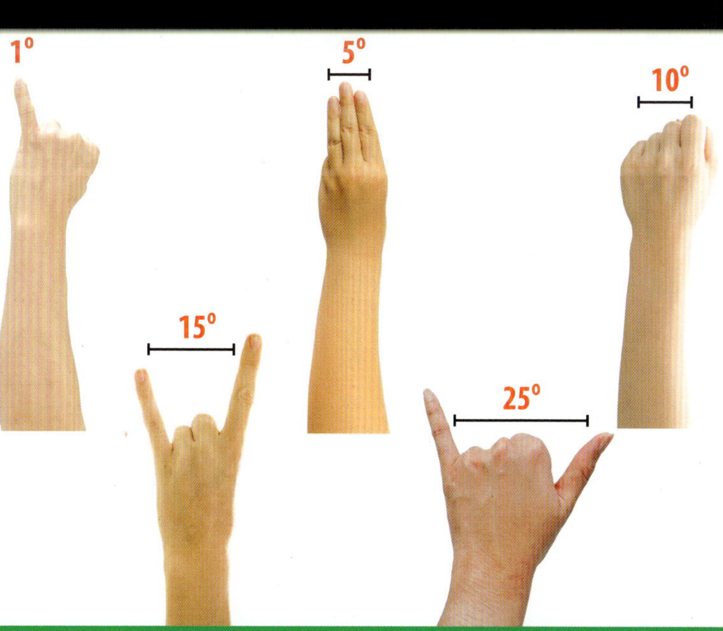

Mapping Stellar Observatories

Stars are most visible from dark areas far away from **light pollution**. Astronomers have created Dark Sky Preserves, Reserves, and Parks in some of the darkest places on Earth. These places give the clearest possible view of the stars. Observatories are often built in these preserves. To get an even closer look at the stars, many observatories are built on mountaintops.

Mauna Kea Observatory, Hawai'i, United States

Mauna Kea Observatory is located on top of Hawai'i's tallest volcano, Mauna Kea, 13,796 feet (4,205 meters) above sea level. The first telescope on Mauna Kea was built in 1970. Since then, more telescopes have been added to the summit. These telescopes took the first clear pictures of planets orbiting a star in another solar system. Scientists also used them to discover a black hole in the **Milky Way**.

Roque de los Muchachos Observatory, Spain

Roque de los Muchachos is located on a 7,860-foot (2,396-m) high mountain on the island of La Palma, in the Canary Islands. The observatory's telescopes include the Gran Telescopio Canarias (GTC). The GTC is the largest telescope of its kind in the world. It uses mirrors that work together to reflect and capture light from space. The telescope has helped astronomers understand how black holes are created when stars die.

Pacific Ocean

Europe

Asia

Africa

Indian Ocean

Iranian National Observatory, Iran

The Iranian National Observatory (INO) began operating in 2018. It was built on top of Mount Gargash at an altitude of 11,800 feet (3,597 m). There are three telescopes at the INO. One operates automatically and posts its observations live online. Another is made up of three lenses designed to see faint objects in distant space. The third telescope, INO340, is designed to look at stars and their life cycles.

Paranal Observatory, Chile

Paranal Observatory is located in Chile's Atacama Desert. It sits on top of Cerro Paranal at an **altitude** of 8,645 feet (2,635 m). The observatory uses the Very Large Telescope (VLT). The VLT is actually four telescopes that work together to look at large portions of the sky. Light enters the telescopes and is sent through a series of mirrors in underground tunnels. The mirrors help focus the images captured by the telescopes. This gives astronomers clear, detailed views of distant stars and planets.

Constellations

Since ancient times, people have grouped stars together. Stargazers imagine they see people, animals, and other objects in the sky. These shapes are called constellations.

Smaller groups of stars are called asterisms. Asterisms are easy to find because they are made up of fewer stars than constellations. Finding an asterism first makes it easier to find the constellation that holds it.

Constellations and asterisms can be used as a map in the sky. The last star in the handle of the Little Dipper, an asterism in the constellation Ursa Minor, is called Polaris. It is also known as the North Star because it is located directly above Earth's North Pole. By looking at Polaris, people in the Northern **Hemisphere** can figure out which direction they are heading.

The **smallest** constellation, **Crux**, takes up only **0.17 percent** of the sky. The **largest**, **Hydra**, covers **3.16 percent** of the sky.

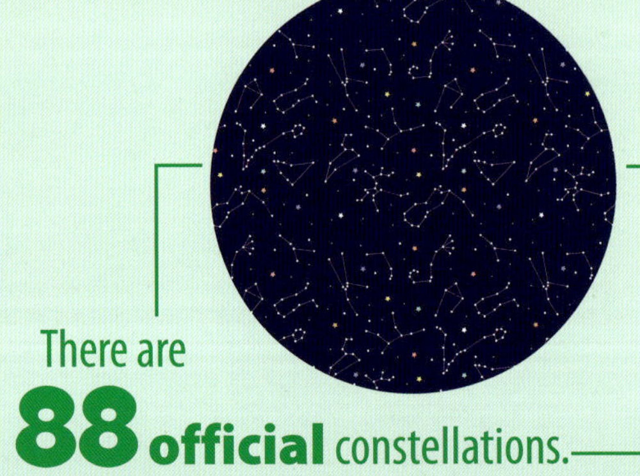

There are **88 official** constellations.

An easy way to find Polaris is to find the Big Dipper. The two stars at the end of the dipper's bowl make a line that points to Polaris.

Stars and Seasons

Not all stars and constellations can be seen from every part of Earth. Typically, only people in the Northern Hemisphere can see Polaris and Ursa Minor. Other stars and constellations can only be seen in certain areas during certain seasons. Crux can be seen from the Southern Hemisphere all year, but it can be seen from some parts of the Northern Hemisphere for a few weeks. The movement of stars and constellations is like a **cosmic** calendar. In the Northern Hemisphere, the constellation Orion is only visible in the winter. Scorpius, the scorpion constellation, is only visible during the summer. Leo, the lion, can be seen in the spring. Pegasus, the winged horse, appears in the fall.

The movement of constellations happens because of something called sidereal time. Normal clocks are based on Earth taking 24 hours to make a complete rotation. This is known as solar time and comes from watching the Sun rise and set. However, if someone watches the rise and fall of stars, Earth seems to rotate faster. It makes a sidereal rotation in 23 hours, 56 minutes, and 4.1 seconds. You can see this in action by standing on the same spot each night and looking at the same star. It will arrive at the same place four minutes earlier each night.

Orion is known for its many bright stars, including Rigel, the seventh-brightest star in the sky.

Sidereal Time

Sidereal clocks are designed to keep time by the stars. Instead of reaching 12 at midnight, a sidereal clock arrives at zero hour 3 minutes and 55.9 seconds before midnight on the first day of the year. It will arrive at zero hour about four minutes earlier each night.

One sidereal year contains about 366.25 sidereal days. On the first day of the next sidereal year, a sidereal clock will start at four minutes before midnight again. A star that was in a certain spot in the sky one year earlier will be back in the exact same spot.

Sidereal and Solar Days

1. It is noon at the base of the arrow. The Sun appears high in the sky.
2. Earth has completed a full sidereal rotation. This means that a sidereal day has passed. However, the Sun does not appear in the same position when viewed from the arrow's base as it did at step 1.
3. Earth has rotated for about four more minutes. A solar day has passed. Now, the Sun appears to be in the same spot in the sky from the arrow's base as it did during step 1.

3 Earth

2 Earth

Sun

1 Earth

Stellar Motion Chart

Track a star in sidereal time as it moves through the sky.

Materials
Graph paper
Pen or pencil

1. Using your pen or pencil, draw a circle on your graph paper. Draw a **horizontal** line through the middle. The circle is a picture of the night sky. The line is the **horizon**.

2. At night, find a bright star to track. Look for a star that will be easy to find each night. Try to pick a star that is part of a constellation. This will make it easier to find in the patterns of the sky.

3. Make a note of the exact spot where you are standing. That is the middle of the horizon line on your paper. Face north or south, making sure that you can see the star. Based on where you are standing, mark east and west at the ends of the horizon line. Then, draw the star on your paper. Place it where it appears in the night sky compared to where you are standing. It should be above the horizon line. If it is directly above your head, it should be at the top middle of the circle.

4. Each night, stand in the exact same spot, facing the same direction. Find the same star at the same time. Mark the star's location on your paper. It should be slightly further west than the night before. Eventually, the star may disappear below the horizon. Check back in about six months, and it may reappear above the horizon on the other side of your circle. After 366 days, it should be in the same place it started.

5. If you connect all the marks you drew on the page, you will see that star's sidereal path through the sky.

Quiz

1 How many stars can be seen with the naked eye?

A: 9,096

2 What was the brightest stellar event ever recorded?

A: SN1006

3 Who discovered proper motion of stars?

A: Edmond Halley

4 What do astronomers call a burning ball of gas that rises out of a star's surface?

A: Granule

5 What is the name of the earliest stage of a star's life cycle?

A: Protostar

6 What telescope was launched into space in 1990?

A: Hubble Space Telescope

7 The Little Dipper is an asterism in which constellation?

A: Ursa Minor

8 How long is a sidereal day?

A: 23 hours and 56 minutes

Key Words

altitude: the height of something compared to sea level

astronomers: scientists who study space

black hole: a large area in space with such strong gravity that light cannot get out

constellations: groups of stars in a pattern

cosmic: having to do with space

fusion: energy created by the joining of atoms

hemisphere: one half of a sphere-shaped object such as a planet

horizon: the line where the ground seems to meet the sky

horizontal: from side to side

light pollution: light that makes it difficult to see things in the sky

light years: the distance light travels in one year; about 5.88 trillion miles

mass: the amount of matter in an object

Milky Way: the galaxy that contains Earth

naked eye: viewed without the help of tools such as telescopes or binoculars

navigate: find the way somewhere

nebula: a group of stars that look like a cloud

nuclear: energy created when atoms are split

orbit: the path one object in space takes around another

parallax: change in the apparent position of something compared to other objects

stellar: having to do with stars

supernova: the explosion of a star when it dies

Index

LIGHTBOX

SUPPLEMENTARY RESOURCES

Click on the plus icon found in the bottom left corner of each spread to open additional teacher resources.

- Download and print the book's quizzes and activities
- Access curriculum correlations
- Explore additional web applications that enhance the Lightbox experience

LIGHTBOX DIGITAL TITLES

Packed full of integrated media

VIDEOS

INTERACTIVE MAPS

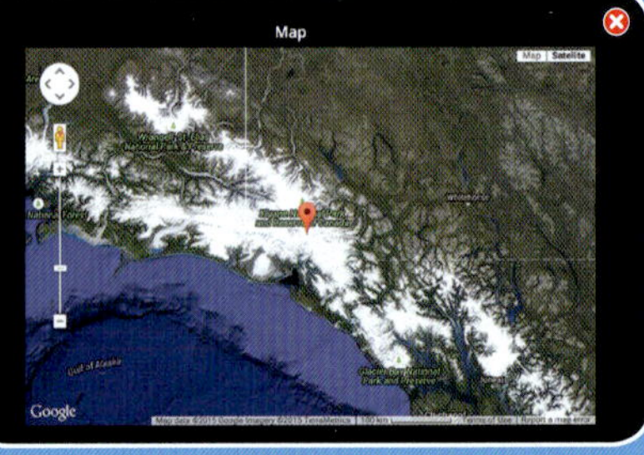

WEBLINKS

SLIDESHOWS

QUIZZES

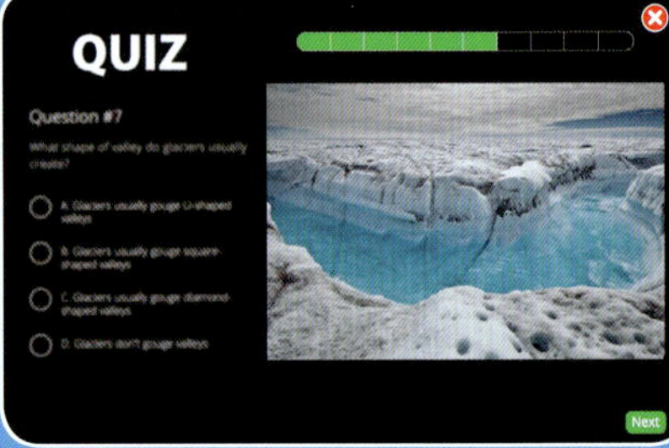

OPTIMIZED FOR

- ✓ TABLETS
- ✓ WHITEBOARDS
- ✓ COMPUTERS
- ✓ AND MUCH MORE!

Published by Smartbook Media Inc.
14 Penn Plaza, 9th Floor New York, NY 10122
Website: www.openlightbox.com

Library of Congress Cataloging-in-Publication Data

Names: Wiseman, Blaine, author.
Title: Earth and the stars / Blaine Wiseman.
Description: New York, NY : Smartbook Media/Lightbox, [2021] | Series: Space systems: stars and the solar system | Includes index. | Audience: Ages 9-14 | Audience: Grades 4-6
Identifiers: LCCN 2020023106 (print) | LCCN 2020023107 (ebook) | ISBN 9781510554702 (library binding) | ISBN 9781510554719
Subjects: LCSH: Stars--Juvenile literature.
Classification: LCC QB801.7 .W458 2021 (print) | LCC QB801.7 (ebook) | DDC 523.8--dc23
LC record available at https://lccn.loc.gov/2020023106
LC ebook record available at https://lccn.loc.gov/2020023107

Printed in Guangzhou, China
1 2 3 4 5 6 7 8 9 0 24 23 22 21 20

062020
111019

Project Coordinator John Willis **Designer** Ana María Vidal

Photo Credits
Every reasonable effort has been made to trace ownership and to obtain permission to reprint copyright material. The publisher would be pleased to have any errors or omissions brought to its attention so that they may be corrected in subsequent printings. The publisher acknowledges Alamy, Getty Images, iStock, Shutterstock, and Wikimedia as its primary image suppliers for this title.